World of science

PHYSICS AND MECHANICS

BAY BOOKS LONDON & SYDNEY

1980 Published by Bay Books
157–167 Bayswater Road, Rushcutters
Bay NSW 2011 Australia
© 1980 Bay Books
National Library of Australia
Card Number and ISBN 0 85835 27 2 9
Design: Sackville Design Group
Printed by Tien Wah Press, Singapore.

FORCES AND LAWS

From earliest times human beings have tried to make machines that would make work easier than having to rely on muscle power alone. As a result, the wheel has enabled us to move quickly and effortlessly from place to place; we have learned how to overcome gravity and fly; we can make clocks that tell us the time of day; we can measure the size and weight of the things around us; and we can look into the smallest units of the matter that makes up our bodies and the world around us.

Mechanics

Mechanics is the science that studies the action of forces on *bodies* or objects. It is concerned with such things as the motion of engines, the flight of aeroplanes, the stresses on bridges and the forces acting on ships at sea.

Under the influence of forces, anything may be either in motion or at rest. A plane on the ground is subject to the force of gravity, the pressure of the atmosphere on its surface and the force of the wind blowing against it. A plane in flight is still subject to the force of gravity, but can now fall, whereas on the ground it could not. It is also subject to the force of air rushing past it as it flies, a much greater force than the wind blowing against it on the ground.

The basic laws of mechanics were laid down by two great scientists, Galileo and Newton, over two hundred years ago. They remained virtually unchanged until the appearance of Einstein's *theory* of *relativity* at the beginning of this century. Even so, many of the basic laws still hold good.

In this painting, Galileo, one of the first modern scientists, is demonstrating his new telescope.

To fly, the plane must lift itself off the ground against the Earth's gravity. The thrust to take off comes either from a jet exhaust or a propeller. The streamlined shape of the plane reduces the air resistance to thrust, and, as the aircraft takes off, lift is provided by the flow of air over the specially curved wings.

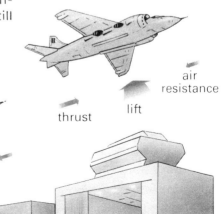

flight

air resistance

thrust   lift

take off

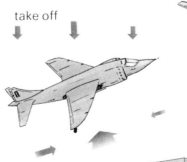

at rest
gravity

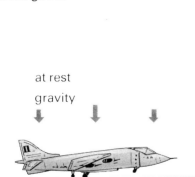

Two great scientists: Albert Einstein (above and top left) and Sir Isaac Newton (top right). Newton's great scientfic contributions were his basic laws of motion and universal law of gravitation, whereas Einstein developed the special and general theories of relativity.

The history of physics

Physics, which incorporates mechanics, is the experimental science that tries to find fundamental laws which will describe matter and energy and what they are and how they work.

At the time of the Greeks, physics was studied together with biology, chemistry, and astronomy under the common name of 'natural philosophy'. The Greeks felt that general ideas and laws were very important, but they based most of their laws on logical reasoning and common sense. So they rarely tested their theories with experiments. Although this led to some strange ideas, many of their theories were correct. Pythagoras, who was well known for his work in geometry, knew that the Earth was a sphere and Democritus taught that matter was made of tiny indivisible particles called atoms. They were not generally believed until two thousand years later. The opposite happened to Aristotle. He thought that the Earth was the centre of the universe and that the sun and the stars moved around it. People didn't believe Pythagoras, who was right and they did believe Aristotle, who was wrong.

After the work of the Greeks, very little progress was made in Europe in the knowledge of physics for several hundred years. Leonardo da Vinci, who was a great painter as well as a scientist, studied mechanics and produced drawings for such inventions as aeroplanes and submarines. In the mid-1500s, Copernicus published his theory about the paths of the planets. He believed that the Earth and other planets moved in an orbit around the sun. This was quite the opposite of Aristotle's view and caused argument among scientists of those times.

The beginning of modern physics

But physics was still a matter of theories and ideas until about 1600 when men like Galileo began to back up their ideas with experiments. Galileo is said to have dropped light and heavy weights from the Leaning Tower of Pisa to prove that all objects fall with the same speed and acceleration. Galileo also produced the first really practical telescope and used it to provide evidence for Copernicus' view that the Earth revolved around the sun. His opponents refused even to look through his telescope and for a long time he was prevented from writing and making further observations.

Below: The leaning tower of Pisa from which, according to legend, Galileo dropped a small and a large weight to demonstrate that all bodies fall to the ground with the same acceleration. Needless to say, they reached the ground together. From this, he was able to calculate that the distance travelled increases as the square of the time taken (diagram, right).

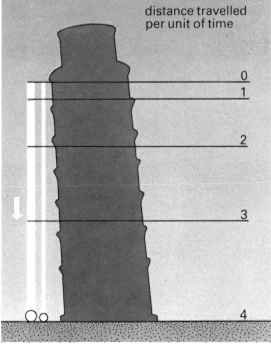

distance travelled per unit of time

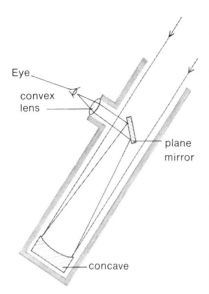

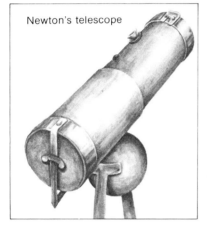

Above: Newton's telescope and a cross-sectional view through it. Parallel beams of light are reflected by a concave mirror on to a smaller plane mirror, which directs the rays out through a convex lens.

Right: There is an equal and opposite force (resistance) to the weight resting on top of the table, which prevents it from moving.

At about the same time as the telescope, around 1600, the microscope was invented. This instrument allowed scientists to study the world of tiny animals and objects. As scientists began to study objects under the microscope they found they wanted to look closer and closer at nature, and, they wanted to be able to make more accurate measurements.

Towards the end of the 1600s, another great scientist, Issac Newton, put forward his *laws of motion* and the law of gravity. Most of Newton's ideas were accepted throughout the scientific world until about 1900 when new techniques for experiment and measurement were developed. Though Newton was right in many of his ideas, he lacked the equipment to carry out the necessary experiments to develop them further.

FORCE, WORK AND WEIGHT

Force and work

In mechanics, *a force* is that which acts on an object to produce a change in its *state of rest* or *motion.* If you push against a table you are exerting a force through the power of your muscles. If you push hard enough, the table will move. That is, it will no longer be at *rest*, it will be in *motion.* If somebody stronger than you pushes against the other side after you have got the table to move, another force will cause the table to stop moving and again be at rest. When a force causes an object to move a certain distance, we say it has done *work* and this can be measured.

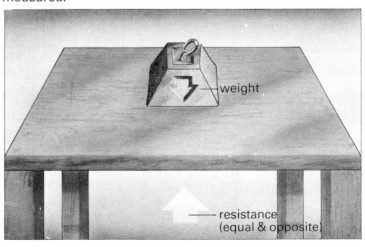

Weight and mass

Another type of force is *weight*. It is the force exerted by gravity on an object or person. For example, a man with a weight of 60 kg is pulled to the ground by gravity with a force of 60 kg. But if he were on the moon, where the force of gravity is one-sixth that of the Earth's, he would weigh only 10 kg, because the force pulling him to the ground would be only one-sixth of that on Earth. That is why astronauts can work for hours on the moon carrying large back-packs. They have only a sixth of their weight on the moon.

In thinking of weight, we have also to think of *mass*, which does not change. The mass of anything is the quantity of matter that the object has. The man's mass remains the same wherever he is, but his weight depends

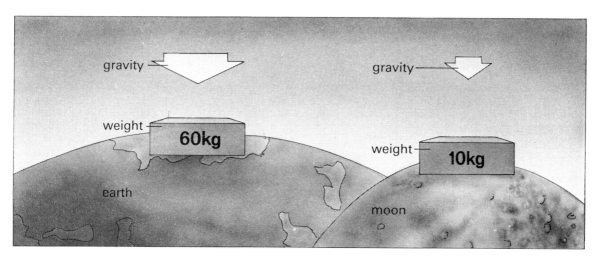

Above: A weight of 60kg is pulled to the Earth by gravity with a force of 60kg. On the moon, the force of gravity is one-sixth of the Earth's, thus the weight would weigh only 10kg.

Left: This beam balance uses known weights in one scale to measure the unknown weights in the other. Any change in gravity will affect both weights equally, enabling the beam balance to be used anywhere.

on the force of gravity. The man's mass would change if, for example, he went on a diet. He would lose mass, and at the same time lose weight. If his mass came down to 54kg, he would now weight 54kg on Earth and 9 kg on the moon.

A spring balance, such as a butcher's scales, measures the weight of a mass because the spring is pulled down by the mass. The greater the mass, the greater the force of gravity and the more the spring will be extended. A set of scales measures differently. It measures mass against mass and is independent of gravity. If you placed two oranges with the same mass in the opposite pans of a set of scales, they would balance exactly the same on Earth as on the moon.

Gravity

The Earth's gravity that causes our weight is referred to as *g* and its effect on a falling body is measured as 9.80665 *metres per second per second*. This means an object dropped from the top of a tall building falls faster towards the ground the further it travels. Under the influence of gravity it *accelerates* as it falls. Gravity is working on our own bodies all the time, as we know only too well when we fall over. It is the force of gravity that makes us fall, and the force exerted by our muscles that prevents us from falling when we are standing up.

One of Newton's Laws states that an object will continue to move in the same direction at the same speed unless affected by a force. A sky-diver leaping out of an aeroplane will continue to drop towards the ground until he opens his parachute because the force of gravity acting on him will cause his speed of fall to increase until the force of the air against his parachute acts against the force of gravity to slow his fall.

The speed and distance travelled by a body falling

Acceleration under gravity is illustrated by this simple diagram. The time taken for a pendulum's swing is dependent on the length of the pendulum, irrespective of the weight of the bob or the degree of swing.

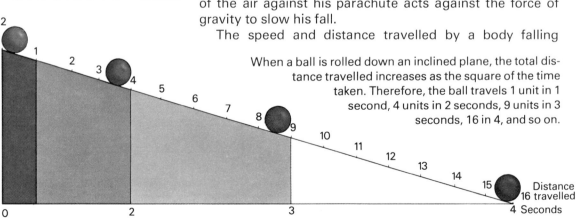

When a ball is rolled down an inclined plane, the total distance travelled increases as the square of the time taken. Therefore, the ball travels 1 unit in 1 second, 4 units in 2 seconds, 9 units in 3 seconds, 16 in 4, and so on.

Above: The space capsule remains orbiting the Earth on a predetermined course because of the planet's gravitational attraction.

Top: This astronaut is walking in space outside the American Gemini 4 space capsule, and he is subject to the same gravitational pull as the capsule.

Left: When a space rocket is launched, the thrust is so great that the rocket can overcome the Earth's gravitational force and can send a space capsule into orbit around the planet.

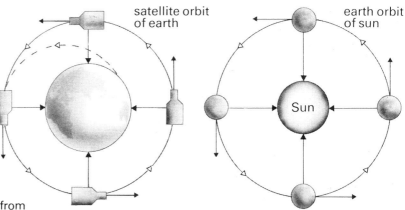

A satellite circles the Earth in the same way as the Earth circles the Sun. Both are held in their respective orbits by gravitational attraction.

Below: A sky-diver can fall from a great height by using a parachute. When jumping from the plane, he pulls a rip-cord which releases and opens the nylon canopy. This acts as a break to slow down the fall of the sky-diver.

under the influence of gravity may be calculated by the equation $y = 4 \cdot 9t^2$, where y equals the distance fallen in metres and t equals the time in seconds since the start of the fall.

Friction

Friction is the resistance to motion of one body caused by contact with another body. The most common form of friction is *mechanical friction*. No surface is absolutely smooth, even a polished metal surface when seen under a microscope shows what look like miniature mountain ranges and valleys. When two surfaces are put together, some of these ranges and valleys interlock and resist sliding until the force applied is large enough to overcome the interlocking effect. A good idea of this may be obtained with two pieces of corrugated cardboard. Matching the corrugations together will give an exaggerated example of the effect of friction and it will need a powerful force to make the two pieces slide over one another.

Another common effect of friction occurs when a hard object forces a soft object to take up a matching shape. This happens when the rubbers of bicycle brakes are forced against the rims of the wheels. The mountains of the hard material form valleys in the soft material and valleys in the hard material are filled by parts of the soft material. In this case there can be a perfect match between the two surfaces and the force required to overcome the friction may have to be extremely large.

Friction can be reduced by *lubrication.* The lubricant forms a thin layer between the two surfaces and instead of there being surface-to-surface contact there is now surface-to-lubricant-to-surface contact. This has the effect of separating the mountains and valleys so that they interlock to a lesser degree. Lubricants can be solids, liquids or gases. The most common lubricant is oil, such as is used in motor cars. Another way of helping to overcome friction is by the use of ball bearings, because frictional forces between rolling systems are much less than between sliding ones. Ships have to overcome a lot of friction as they pass through the water; the hovercraft uses a cushion of air to get the craft above water, and

Below: Friction is the force that tends to stop movement between the weight and the top of the table. However, when a film of oil is placed between the two surfaces the friction is reduced, and even a heavy weight can easily be moved.

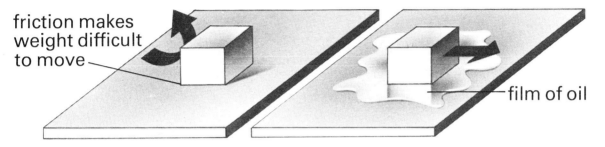

a hydrofoil raises itself almost out of the water on its foils, which offer less resistance to the water than does the hull.

The winds that blow across the earth are slowed down by friction as they pass over the mountains and valleys of the planet. Above a height of about 450 metres, winds can be blowing at twice their speed at ground level.

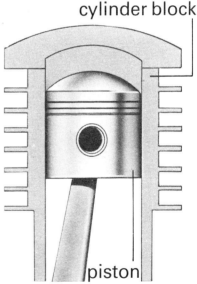

Above: A thin film of oil can reduce friction in a cylinder by separating the cylinder from the surface of the piston.

Left: A hydrofoil can 'ski' over the surface of the water on foils, which are attached to struts on either side of the hull.

Above: Centrifuges can be used to purify and separate mixtures. When the handle is turned, the substances inside the tubes separate into layers of different densities as the test tubes are whirled around in circles.

Below: This boy, skateboarding at speed, can keep his balance by leaning his body the opposite way to the direction in which it is being automatically thrown.

Inertia

Before a force can move a body it must overcome not only friction but inertia. This is the resistance to movement caused by the weight of a stationary body. Even if there were no friction, this resistance must still be overcome before the body can move. The force that is needed to overcome inertia is greater than is needed to keep it moving, which is why even a heavily laden truck can increase speed when moving uphill from a standing start.

Centrifugal and centripetal force

These two forces act on moving bodies. If you tie a weight to a piece of string and whirl it around, you can feel the weight wanting to pull itself away from the string. The faster you whirl, the more the weight seems to tug. If the string breaks or you let go, the weight flies off in a straight line. The force that causes the weight to tug outwards is called *centrifugal force.* The force that stops the weight flying out and makes it move in a circle, is supplied through the string and your hand. This is called *centripetal force.*

Many of the exciting rides at fun fairs use these two

forces. One that is often used is called a *centrifuge*. This is a huge drum where people walk inside and stand against the wall. When the drum is turned at fairly high speed by a motor the people inside stick to the walls like flies on flypaper. What causes them to stick is centrifugal force. Like the weight at the end of the string, they want to fly outwards but the wall keeps them there.

When a car takes a corner at high speed you are thrown against the door by centrifugal force. If you were not stopped by the sides of the car you would fly out. Skateboarding provides another example, as does snow skiing. If you skateboard or ski at speed around a curved track your body will automatically be thrown outwards and to keep your balance you have to lean your body in the opposite direction. The two forces are equal and opposite. If the weight is too heavy it will build up a lot of momentum and the string will break. If you lean too far inwards you will topple over, if you don't lean far enough you will fall outwards.

Above: This 'round-up ride' at a fair is yet another type of centrifuge. When it revolves the people stick to the sides.

Below: Tie a small weight to a piece of string, and whirl it around in a circle. You will feel the weight tugging outwards – this is centrifugal force. Your hand supplies the centripetal force through the piece of string.

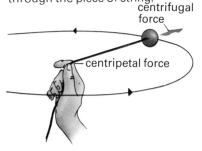

These people are enjoying all the fun of the fair in a 'rotor', which is another type of centrifuge. When the rotor turns, they are literally glued to the walls as the floor drops away underneath. However, as the rotor slows down, they drop on to the floor.

Centrifugal force is the basis of machines called centrifuges, used to separate mixed liquids or solids from liquids. If you stir a bucket of water with sand in it, you will see the grains of sand being thrown to the sides of the bucket. Centrifuges used in biology, medicine and chemistry operate on this principle. A simple type of centrifuge is the spin dryer. Wet clothes go into a drum with holes in it. As the drum spins, centrifugal force causes the water to be thrown out through the holes while the laundry remains in the drum.

SIMPLE MACHINES

A machine is a device that is used to do work or help someone to do work. Six devices are called *simple machines*, though when they are used in certain ways, they may appear to be more complicated. The six devices are: the *lever, wheel and axle, pulley, inclined plane, wedge,* and *screw.*

Mechanical advantage

It is often useful to know exactly what advantage is gained by using a machine. A simple machine gives what is called a *mechanical advantage.* It is measured as the *ratio* of the load moved to the effort applied to move it. For example, if by using a machine you can move a load with half the effort required without it, the mechanical advantage of that particular machine is 2:1. But although your effort is less, the work done on the load is the same because of the machine. Work done can be calculated in the following way; work done equals the load multiplied by the distance it is moved. Work done also equals the effort multiplied by the distance over which it is applied. Thus, if the mechanical advantage is 2:1, the effort in this case must be applied over twice the distance that the load

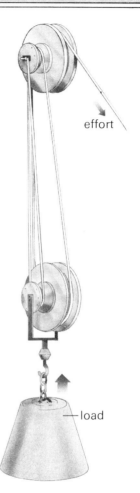

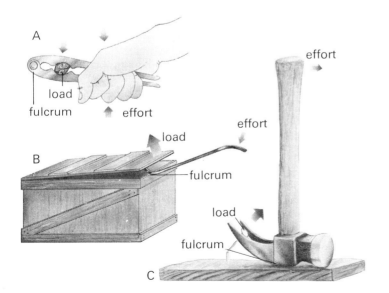

(A) The load is between the fulcrum and the effort in this extremely simple lever.
B) This lever is ideal because the fulcrum is positioned between the effort and the load.
C) The claw hammer can also be used as a lever.

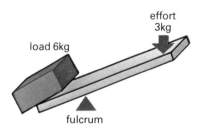

Above: When an effort of 3kg is exerted on one side of this lever, it can lift a load of 6kg, therefore giving it a mechanical advantage of 2.

Below: Three types of levers.
1 First-class lever (pump-handle)
2 Second-class lever (wheelbarrow)
3 Third-class lever (human arm)

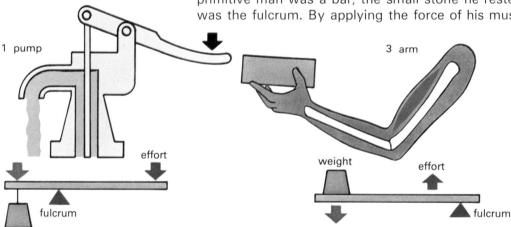

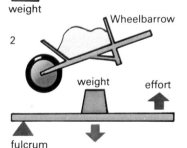

moves. The ratio of the two distances is called the *velocity ratio*, and in a perfect machine the mechanical advantage should equal the velocity ratio.

Work done = Load × distance
Work done = Effort × distance

Unfortunately, the perfect machine does not exist, because no matter what we do, we are always confronted with friction. This means some of the work done in applying the effort is used to overcome friction and so is not used in actually shifting the load. The actual mechanical advantage of the machine is therefore reduced. That doesn't stop us using machines, however, and we calculate the efficiency of a machine as the ratio of the *mechanical advantage* to the *velocity ratio*.

The lever

At its simplest, the lever is a rigid bar pivoted on a fixed support or *fulcrum*. The tree branch perhaps used by primitive man was a bar, the small stone he rested it on was the fulcrum. By applying the force of his muscles at one end of the branch he could use the lever to move a large stone.

The basic law of the lever states that the product of the effort and the distance between the fulcrum and the point at which the effort is applied is equal to the product of the load and the distance the load is from the fulcrum.

The law of the lever leaves out friction, but in a lever the effect of friction is quite small and for all practical purposes doesn't count.

Levers are of different kinds and the three main classes of lever depend on where the fulcrum is placed. In a first-class lever, the fulcrum is between the effort and the load. A crow-bar is a first-class lever, so is a see-saw. The law of the lever can be demonstrated with a simple type of see-saw, such as a plank of wood across a log. If you put the plank so that equal lengths are on each side of the log, then two people of the same weight will be balanced. But if the fulcrum is not in the middle, you need a heavier person on the shorter end.

A second-class lever has the fulcrum at one end and the load between the fulcrum and the effort. Nutcrackers are an example of this type and when you have a tough nut to crack you will notice how very efficient this type of lever can be. A wheelbarrow, which of course uses both a lever and a wheel to help a man move a load heavier than he can carry, also uses this type of lever. The third-class lever also has the fulcrum at one end, but this time the effort is between the fulcrum and the load, an example of a third-class lever is a fishing rod. This is not as efficient a lever as the other two in terms of mechanical advantage.

The wheel and axle

While the wheel and axle are what make our transport move along, the principle is also used to do a great deal of our work for us in other ways. A centre-pin fishing reel is an example, as is the hand *capstan* used to haul in a ship's anchor. In the capstan, a rope carrying the load is wound around a small diameter axle. Effort is applied around the

Above: This English water-wheel, which turns a millstone, is driven by natural energy harnessed from a nearby waterfall.

Below: How the water-wheel works. The force of the water, flowing beneath the wheel, turns it. The torque, or turning effect, is produced by the force acting near the wheel's edge.

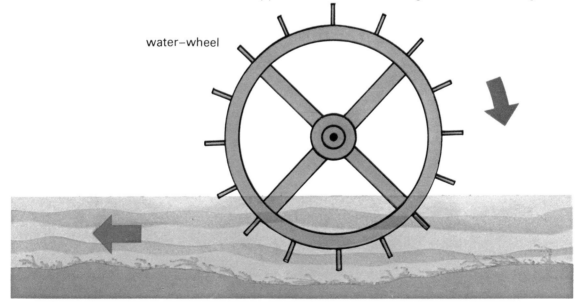

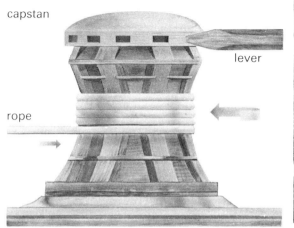
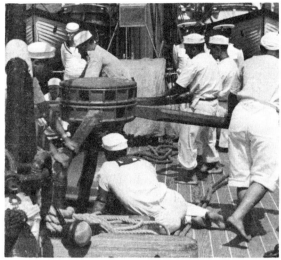

Above: In the simple capstan, power is applied manually to the lever, causing the whole capstan to revolve and thus the rope is wound in.

Above right: Most capstans are now mechanically or electrically driven, but these naval cadets competing in the Tall Ships Race have to use their muscle power. Capstans were once used to raise ships' anchors and hoist cargo abroad.

The windlass is used to raise a bucket containing water from a well-shaft. When the handle is turned, the rope is wound in and the bucket rises.

windlass

circumference by the use of long levers connected to one end of the axle. The mechanical advantage is obtained from the difference in the two diameters. It is not always necessary to have an actual wheel, so long as a force can be applied which would be the same as if a wheel were to be used. For example, a simple well may have a bucket hanging from a rope wound around a drum at the top of the well, but instead of a wheel there is a handle in the shape of a crank. When you turn the handle of the crank it is the same as if you were turning a wheel.

The pulley

The pulley is a variation of the wheel and axle, but instead of the rope being wound around anything, it passes over the pulley, which is mounted on an axle. Pulley systems are used, for example, in cranes and hoists.

Using a single fixed pulley with the rope tied to the load and pulling on the other end doesn't give any mechanical advantage. The same amount of work is still needed to lift the load off the floor. The advantage, however, is that you can alter the direction of effort, you can stand back and pull instead of having to bend over and lift.

But if you add another pulley that can move and one end of the rope is attached to a support you have a simple *block and tackle* which gives a mechanical advantage of 2:1. The support to which one end of the rope is fixed takes half the weight of the load. More pulleys can be added to the system, each increasing the mechanical advantage and convenience of lifting. Motor mechanics use a system of pulleys to lift motors out of cars and it seems they do hardly any work at all. A chain is used instead of a rope, or on a heavy crane, wire ropes are used. In a crane the principle of the capstan is used horizontally to increase lift and decrease effort. This horizontal drive or axle on which the rope is wound is a windlass or winch. The jib of the crane is used to suspend the hook over the load so that it can be pulled straight up, and systems of levers are sometimes combined into the construction of the jib.

Pulleys can be used to lift the engines out of these Ford rally cars. Using a chain pulley, hardly any physical effort is needed from the mechanic.

By using a wooden ramp, these two men can easily load a heavy packing case on to the back of their removal truck. Using an inclined plane, less force is necessary although the distance moved is increased.

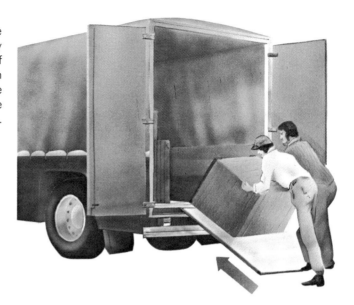

The inclined plane

The inclined plane helps overcome the problem involved in raising heavy objects against the force of gravity. A simple example is the wooden ramp placed against the back of a truck. A heavy packing case which would be very difficult to lift can be pushed up the board. The gentler the slope the greater the mechanical advantage but the further the packing case must be pushed. The amount of effort required is increased by friction between the packing case and the board and this can be reduced by using a smooth surface or rollers. Ancient ship builders used rollers to get their ships to and from the sea, and this system must still sometimes be used today. Furniture removal vans often have built-in ramps with rollers to make loading easier.

The screw

Opposite: Boats, such as this lifeboat, are still launched down an inclined plane. The Royal National Lifeboat Institute answers calls from any ships in distress.

The screw is actually a variation of the inclined plane in the shape of a spiral around a central core. A turning effort applied to the head of the screw becomes movement of the screw along its axis. The screw has a *pitch*, which is the distance between two successive crests of the screw thread. Turning the head of the screw once makes the screw advance by this distance. A common example of a screw is the car jack. When the jack is placed under a car

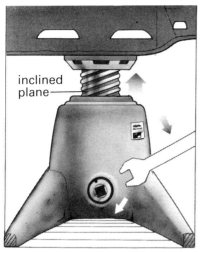

Above: In the screw jack, the screw thread acts as an inclined plane. The base screw can be turned with a spanner with the minimum of effort.
Above right: Close-up view of a ratchet-type car jack.
Opposite: This unusual picture of a gigantic spiral staircase can help us to understand how a screw works.
Below: The splitting force of the wedge is 6 x that of the blow of a hammer.

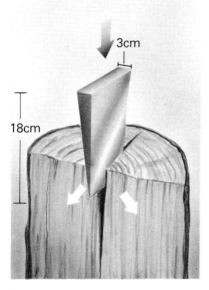

it is turned by means of a long handle which acts as a lever in a capstan. The combination of the mechanical advantages provided by the lever and the screw makes it quite easy to lift a vehicle weighing as much as a tonne. If you use a jack of this type you will notice that, although very little effort is required, the effort has to be applied over a long distance, for the jack handle has to be turned many times.

The wedge

Wedges perform a splitting action in timber or stone. When effort like the blow of a sledgehammer, for example, is applied to the thick part of the wedge, the force is transferred to the angled faces. The force is then transmitted to the timber or stone, causing it to split. An axe is a wedge, so is a nail. If you try to hammer a thick nail into a thin piece of wood you run the risk of splitting the wood as the fibres are wedged apart.

Applications of simple machines

A flagpole uses a pulley to make it possible to get the flag to the top of the pole without having to climb up; sailing boats use pulleys to help control the sails and the fan belt on a car must use a pulley to apply some of the work of the engine to drive the generator.
　　A revolving typist's chair and a revolving piano stool in some cases use a screw to enable you to raise or lower

the height. A spiral staircase is a form of screw and helps us to understand the screw as an inclined plane in the shape of a spiral. Climbing a spiral stair certainly makes us travel a greater distance to climb a height than if we went straight up, but it is also certainly much easier.

A winding mountain road is an inclined plane, as are the ramps built into many supermarkets which make it easier to push a loaded trolley or a baby carriage from one level to the next. A flight of steps is a variation of the inclined plane.

Gears

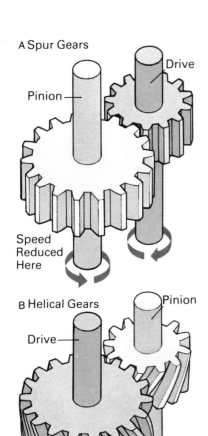

The toothed wheels used in mechanical toys, clocks and machinery are gears. They operate in pairs to transmit the movement of one rotating shaft to another. They are used to speed up or slow down the movement, to transfer action from one spot to another, or even to change the direction of the movement.

Generally, one wheel is larger than the other and the difference in size between them determines the result. The driving wheel is usually called the *gear* and the driven one the *pinion.* When the smaller wheel is the one doing the driving, then the system acts as a *speed reducer.* When the larger wheel, or gear, is doing the driving, then the system acts as a *speed increaser.* For example, a 30-tooth gear driving a 15-tooth pinion will double the speed. If a 15-tooth pinion drives a 30-tooth gear, the speed will be halved.

When the gear and pinion mesh, they turn in opposite directions. If they are required to turn in the same direction another toothed wheel called an *idler* is positioned between the gear and the pinion. The gear turns the idler in the opposite direction and the idler turns the pinion in the opposite direction to it, which becomes the same

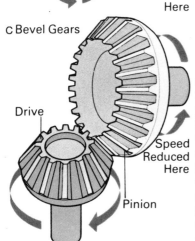
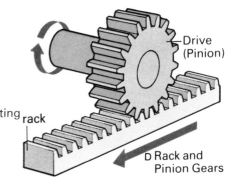

A Spur gears have straight teeth and can be at least 95 per cent efficient.
B Helical gears have spiral teeth and they generate an axial thrust.
C Bevel gears have straight teeth cut in an angle, connecting shafts whose axes intersect.
D Rack and pinion gears - when the pinion is turned, it meshes with the rack.

1. Yoke lock
2. Selector cam
3. Selector Yoke 3rd/4th Gear
4. Selector Yoke 1st/2nd Gear
5. Selector Yoke Reverse Gear
6. Selector Rod
7. Gear Stick
8. Transmission Shaft
9. Idler Gear and Shaft for Reverse Gear
10. Reverse Gear Train
11. 1st Gear Train
12. 2nd Gear Train
13. 3rd Gear Train
14. Layshaft
15. Layshaft Gear Train
16. Bearings
17. Clutch Shaft
18. Locking Dogs and Cone Clutch
19. Sliding Dogs
20. Synchromesh Ring
21. Selector Fork

This model 4-speed gearbox is typical of many small saloon cars. The baulk ring synchromesh prevents the driver from engaging a gear until both the road and engine speeds harmonize.

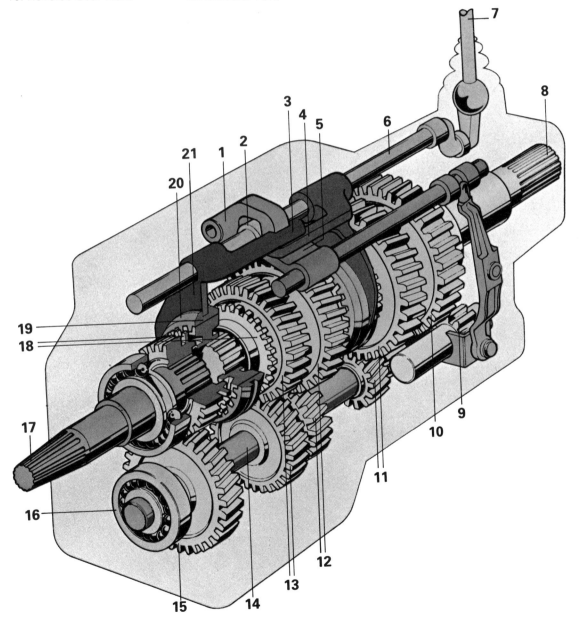

This train that is crossing a bridge in Switzerland has a rack and pinion arrangement of gears. Many trains like this one are still in operation in mountainous areas.

direction as the gear. When the drive shaft and the output shaft are some distance apart, a number of idler gears can be introduced into the system to transfer the movement. This kind of arrangement is called a compound train of gears.

Another type of gear arrangement is the rack and pinion system. This is used, for example, in some car steering systems and in camera and microscope focusing mechanisms. The rack consists of a gear with its teeth set in a row. As the pinion is turned, it meshes with the rack, moving it or moving along it in a straight line.

A gearbox for a motor car is a system of gears and pinions equipped with a device known as a *selector*. This allows a constant speed input to be changed to a varying output by the simple process of bringing together different combinations of gears and pinions.

CONSERVATION OF ENERGY

Energy and work

To carry out any task, to perform any work, we have to use *energy*. The simple machines, such as the block and tackle, enable us to use energy more efficiently to lift weights.

Many forms of energy exist, for energy is necessary for all life. Green plants get energy for growth in the form of heat and light from the sun. Animals get their energy by eating plant and animal food, with the digestion of food providing *chemical energy*. We use chemical energy in other ways, particularly by burning fuels such as coal, gas,

These athletes need to conserve some of their energy for the final burst of speed at the end of an important race. We all need energy to carry out any task or work, however simple.

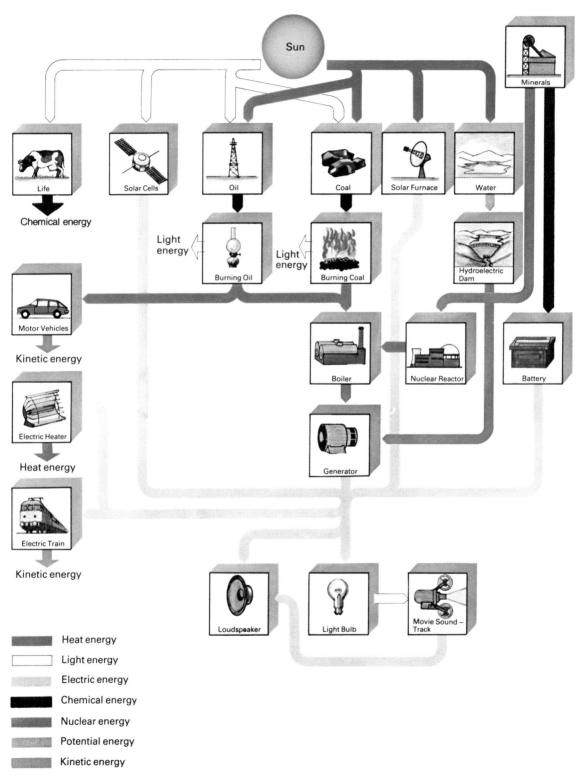

and oil for heating. In this case the chemical energy is converted into *heat energy.*

An object that moves uses energy to do work by moving. The energy of a moving object is called *kinetic energy.* A moving motor car has kinetic energy, as do the moving atoms of matter. The wound-up spring of a clock has energy stored within it and until that energy is released it is called *potential energy.* The waters of a lake high up behind a dam have potential energy which we can use once we direct them through a turbine to produce electricity.

A ball rolling downhill will lose potential energy, but will gain kinetic energy. If the ball rolls partly up an opposite hill, it gains potential energy but loses kinetic energy and it slows down. Friction with the ground will eventually cause the ball to come to rest at the bottom of the hill, unless more energy is applied, such as by kicking it again.

Perpetual motion

Many people have been fascinated by the idea of producing movement that will go on forever without needing any continuing driving force. Even today the idea of *perpetual motion* still fascinates some people.

There are two main reasons why a perpetual motion

Opposite: Energy is essential for all forms of life. On Earth, all energy comes from the Sun and minerals. This chart shows how energy is converted for everyday domestic and industrial use. Although the Sun is the source of all energy, it must pass through plants before we use it as food. It may be stored in coal which releases heat as it burns. This heat can change water into the steam needed to drive a turbine to supply electricity or power.
Left: This mediaeval machine was designed to demonstrate the principle of perpetual motion. However, like later experiments, it was doomed to failure. Friction between the machine's components stops the screw carrying water into the tank.

machine seems impossible to build. One is friction. No matter what materials are used, there is friction and even a very small amount of friction will eventually cause the machine to run down. The other reason is that to produce power, you must use energy. This is a natural law and it is usually called the First Law of Thermodynamics. A so-called perpetual motion machine built in the 1500s failed because it broke this law. It consisted of a wheel with heavy spiral spokes, between and along which heavy balls could move. The idea was that as the balls fell, they would hit the rim of the wheel, causing it to turn. When the balls reached the bottom they would roll towards the hub and eventually reach the top, ready to fall again. But first of all there was friction in the wheel which caused a loss of energy and slowed it down; second, as the balls fell some of their energy was lost when they hit the rim of the wheel.

These electrical technicians are perched high up above the ground on cables. Electrical energy is carried through a grid system on these cables.

Conservation of energy

One of the scientific laws of nature is that energy can neither be created nor destroyed. This is the *law of the conservation of energy*. It means that energy cannot appear from nothing, nor disappear into nothing. Energy never disappears from the universe, although it constantly changes its form. It can only be changed from one kind of energy to another, so that the total amount of energy remains the same. For example, burning fuel raises steam. The steam drives a turbine generator which produces electricity. In this way, the heat energy of fuel is changed into electrical energy. The electrical energy may then be turned back into heat energy in an electric fire or into kinetic energy in an electric motor.

Although energy cannot be destroyed, some of it is always lost, so there is still no hope of a perpetual motion machine. In the case of electricity generation, for example, some of the energy released by burning fuel escapes through the sides of the boiler. In travelling along the electric power lines, some of the energy of the electricity is also lost as heat. This lost heat goes back into the atmosphere where it may heat up something else, or be converted into energy in the form of a warm air current.

Lightning is the visible manifestation of huge sparks of electricity in the air. Sometimes, enormous positive and negative charges build up in thunder clouds, and these are discharged as lightning.

MEASUREMENT

In our daily lives, as well as in science, we need to know the measurements of the things around us. In the days before accurate measuring devices were available, things were measured by other familiar objects, often parts of the body. This is still described as *rule-of-thumb* and you can see that you can use your thumb as a regular measuring tool. Horses today are still spoken of as being so many *hands* high, a hand being the old measure indicated by the width of a man's hand. A *foot* is said to be the length of the foot of one of the kings of England and that distance was marked in a special place so it could be used as a standard of measurement.

Length, breadth and thickness

Accurate measurements between points are taken with various measuring devices marked in standard units, such as millimetres, centimetres and metres. A ruler or *scale* is the commonest measuring instrument for finding distances between points in straight lines. Length, breadth and thickness can all be measured in this way. For accurate measurements in engineering, a *vernier caliper* is often used.

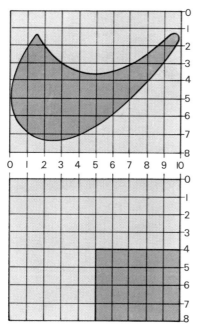

Below: The area of an irregularly shaped object can easily be calculated by tracing the outline on to a piece of squared paper. Just count the number of squares that lie completely within the keyline. Estimate the number of part squares and add the two figures together.

Above: To calculate the area of a square or rectangle, draw the outline on to squared paper and count the number of horizontal and vertical squares. Area = length x breadth, thus the area is 5 x 4 = 20 square units.

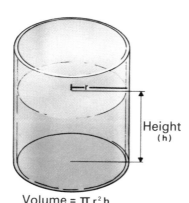

Volume = $\pi r^2 h$

Above: To calculate the volume of a liquid in a cylinder, you can apply a simple formula ($\pi r^2 h$). The volume of the small cube is length x breadth x height (2 x 2 x 2). Therefore the volume is 8 cubic units.

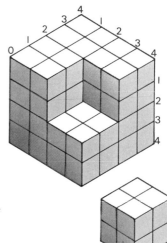

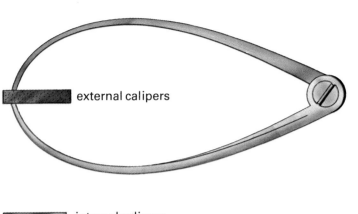
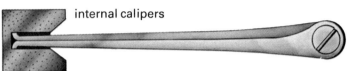
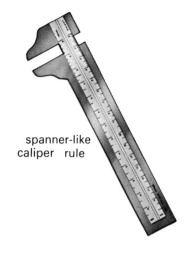

Area

The *area* of a *surface* is the amount of space it occupies. Finding the area of regular objects such as squares, triangles and rectangles is comparatively easy. The area of a table top, for example, is the product of its length and breadth. Finding the area of irregular objects is less easy. You can make an estimate, however, by tracing the object's outline onto a sheet of squared paper and counting the number of squares which are wholly within the outline and estimating the number of part squares. By adding the two together you will arrive at a reasonably accurate measurement of area.

Scientists use calipers to get accurate measurements in cases where a rule would be difficult to use. Internal calipers are used for inside measurement; external calipers are used for outside measurement; and the spanner-like caliper rule may be used for inside and outside measurements.

Volume

Volume is the total amount of space occupied by an object or substance. The substance may be air, or it may be a solid or a liquid. A litre is a measurement of volume and a one litre soft-drink bottle contains a litre of air when empty and a litre of soft-drink when filled.

Weight, mass and density

The mass of an object is the amount of matter (atoms) it contains. The weight of an object is a result of the action of gravity on the mass of an object. The action of gravity is in fact a force. Density refers to how closely packed the matter is in an object. For example, a brick is quite heavy, it would be much heavier still if it were made of lead, for lead is a dense metal; if it were made of wood it would probably be much lighter than an ordinary brick, for wood is not as dense as brick.

Scales or spring balances are used to measure weight. In a pair of scales, weights are used to balance the object being weighed and the total of the weights used gives the weight or mass of the object.

Standard units

It is necessary to be able to measure length, weight or mass, any volume of objects and bodies, as well as distances between points on the Earth's surface and in space, and time. These basic measurements are essential for us to be able to calculate other measures like area, force, pressure and speed.

So that measurements can be compared, standard units are needed, and in this century more and more countries have adopted metric measures and have abandoned their traditional units. This has led to Australia, for example, deciding to abandon the old Imperial or British measures and adopt the international metric system.

An international standard is adopted for each basic unit and dimension and the ways of converting one measure to another are fixed according to agreed values. Thus, the basic metric measure of length is the *metre*, which is

This picture sequence shows the principle of Newton's first law of motion, which states that a force is necessary to make an object accelerate or slow down. **1** The girl pedals hard, or uses a lower gear, to make the bicycle accelerate. **2** She applies the brakes gently and the movement is slowed down by an opposite force. **3** When she squeezes the brakes hard and tries to pedal, the forces are balanced and the bicycle will not move at all.

divided into 1000 millimetres, which are our normal, everyday units of length. The basic measure of weight is the *kilogram*, which contains 1000 grams, and these are our everyday units.

The basic measure of volume is the *litre* or cubic decimetre, which contains 1000 millilitres, which are our everyday units for measuring volumes of liquids and gases.

DYNAMICS AND MOVEMENT

The laws of motion

Dynamics is the study of the causes of *motion* or movement and is understood by using Newton's three laws of motion:
1   All objects remain at rest or move with constant velocity (speed) unless a force acts on them.
2   Acceleration (the rate of change of velocity) of a body is directly proportional to the force acting on it.
3   To every action there is an opposite and equal reaction.

The first law tells us that a force is required to make an object move faster, or to slow it down. If you are riding a bike and you want to go faster, you have to push the pedals harder. On the other hand, to slow down you have to apply a force by means of the brake. Another way to slow down is to stop pedalling on a hill. Here gravity acts as a force to stop your motion.

To understand the second law, it is only necessary to realise that if you pedal slowly your speed will increase slowly, but if you pedal faster you will go faster. If you apply the brakes quickly and hard you will stop fast. If you apply them slowly you will naturally slow down slowly.

A simple illustration of the third law takes place when you push your hand against a table top. As you push your hand down, the table shows an upward reaction which you feel as pressure on your hand. On your bicycle, when you put the brakes on, the force of your action results in an opposite and equal reaction and the machine stops moving.

Velocity

The rate of movement of a body is called *velocity*. Velocity is measured by time and distance, so the velocity of a ball is the distance in metres that it travels in a second, for example, just as the velocity of a motor car or aeroplane is measured in kilometres per hour.

Increase in the speed of movement of a body is called acceleration; when a car speeds up, its passengers experience the effects of acceleration as they are pushed back in their seats. When the brakes are applied, the car slows down or decelerates, and the occupants experience the opposite effect.

Opposite top: This is not a firework but the sparks flying out when a length of metal tubing is cut using the abrasive cutting technique. It is possible to measure the velocity of the sparks as they travel through space.

Opposite bottom: The moving spheres of Newton's Cradle demonstrate the principles of mass, velocity, acceleration and momentum. When the last sphere is struck, it moves and hits the next sphere in line and thus sets up a chain reaction.

Below: A child's stationary spinning top is brightly coloured, but when it is spinning fast, the colours merge, and it appears white. The rate of movement of the spinning top is known as velocity.

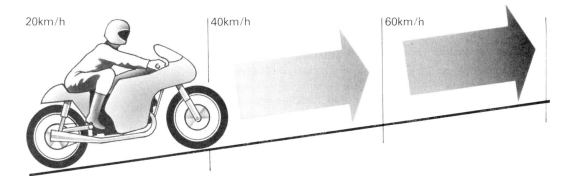

Acceleration

Acceleration is produced by a force acting on an object to change its velocity. If no force is acting on the object, then the object will not move. If the object is already moving and no force is applied to it, it will continue to move at the same speed but there will be no acceleration. Acceleration only applies when the velocity alters.

The greater the mass of the object, the more force needed to make it accelerate. Another way of looking at it is that the greater the mass the less the acceleration produced by a particular force. Heavy trucks have larger engines than most cars, but they still take a long time to get up speed compared to a car. The same applies to deceleration, the larger the mass the more force needed to stop it moving. Trucks are equipped with more powerful brakes to help them stop quickly.

Acceleration is measured as the *rate of increase* of *velocity* and is described as metres per second per second in the same way as the increasing velocity of a body falling towards earth under the influence of gravity.

Above: Acceleration is the rate of increase of velocity. This motorcyclist is accelerating to make his machine go faster. As the motorcycle accelerates, its speed continues to increase.

Opposite: When a space rocket is launched, the fuel burns and gases are blown out of the rocket nozzle, and the rocket moves in the opposite direction. This demonstrates Newton's third law of motion, which states that for any action there is an equal and opposite reaction.

Pressure

When we press our hands on a table, the force we are applying is pressure. It is measured as the force that acts on a particular area. According to Newton's third law of motion, for every action there is an equal and opposite reaction. When we press on the table top, we find that the table exerts an equal pressure upwards, so it does not collapse.

Different substances react differently to pressure. A table is not bent when we press on it because the molecules which make up wood are held together by

This balloonist is wearing a pressurised suit, which will protect him from the effect of reduced air pressure when he is high above the Earth.

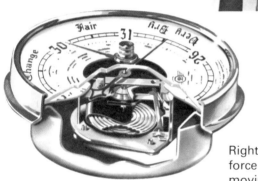

Below: The aneroid barometer can measure atmospheric pressure.

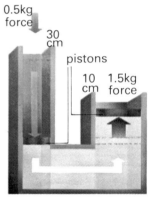

Right: When a pressure of 0.5kg force is exerted on the small piston moving through 30cm, it does the same work as a force of 1.5kg moving through 10cm.

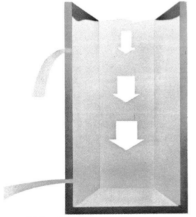

This simple experiment with a tank of water demonstrates that the pressure is greater at the bottom of the tank than near the top.

strong forces which are difficult to break. However, clay is held together by weaker forces, strong enough to overcome gravity, but not strong enough to withstand the pressure we can apply with our hands. Water is held together by even weaker forces and cannot even stand up against gravity. Water will always flow to the lowest possible position. Gases have forces so weak the gas can easily be compressed.

The air around us exerts a pressure on our bodies because of the weight of the air itself. We are used to this pressure and hardly notice it. But the higher up we go, the less is the weight of air and so the less pressure. High-flying aircraft have pressurised cabins in which air is under pressure about the same as we are used to at ground level. The average pressure of the air at ground level is about 14.7 pounds per sq. inch (6.89kPa), or 1 atmosphere.

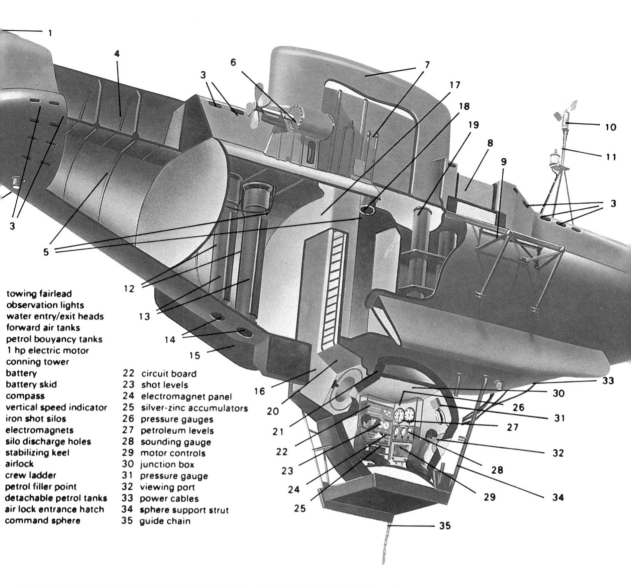

1. towing fairlead
2. observation lights
3. water entry/exit heads
4. forward air tanks
5. petrol bouyancy tanks
6. 1 hp electric motor
7. conning tower
8. battery
9. battery skid
10. compass
11. vertical speed indicator
12. iron shot silos
13. electromagnets
14. silo discharge holes
15. stabilizing keel
16. airlock
17. crew ladder
18. petrol filler point
19. detachable petrol tanks
20. air lock entrance hatch
21. command sphere
22. circuit board
23. shot levels
24. electromagnet panel
25. silver-zinc accumulators
26. pressure gauges
27. petroleum levels
28. sounding gauge
29. motor controls
30. junction box
31. pressure gauge
32. viewing port
33. power cables
34. sphere support strut
35. guide chain

Aerodynamics and flight

This is the science that deals with gases, particularly air, in motion. It also studies the flow of air over stationary bodies, such as the effect all buildings have on the passage of winds. A third study of aerodynamics is the movement of bodies through the air. This is important in the design of aeroplanes and racing cars where, the less the drag from the wind resistance, the faster they can go.

Any body moving through the air experiences a force arising from the *resistance* of the air. This resistance shows up in two ways, at right angles to each other. One, called *lift*, is directed vertically upwards. The other, called

Men can explore the deepest areas of the world's oceans using a bathysphere. Pressure exists in the sea as well as the atmosphere, and the crew inside this electrically powered vehicle can breathe air at normal pressure. As external pressure on the bathysphere increases, the degree of watertightness also grows.

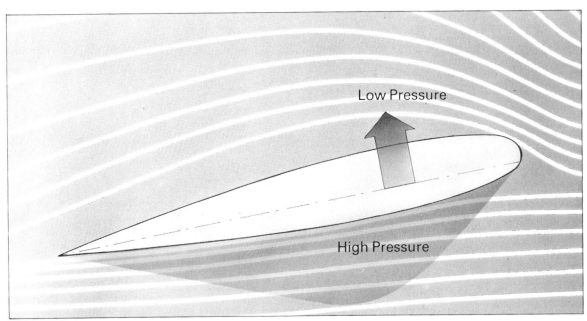

Moving air creates a low pressure area above the wing (aerofoil) and a high pressure area below the wing. Thus lift is produced which enables the aerofoil to rise. Once airborne, more lift can be created by increasing the plane's speed.

drag, acts horizontally and opposite to the direction the object is travelling. For a plane to fly, the *lift* must overcome its *weight*, and the drag must be small enough to allow the plane to move forward.

The basis of a plane's ability to fly is in the wings which have the shape of an *aerofoil*. In cross-section an aerofoil has a rounded nose, a sharply curved upper surface, a flatter under-surface and a sharp tail. The aerofoil is tilted at a slight angle to the direction of the air flow, and lift is achieved in two ways. The air passing over the upper surface of the aerofoil must move faster because of its curved shape, so the pressure of the air below it decreases. At the same time, the air flowing under the flatter surface is slowed down so that its pressure increases. These two forces produce *lift*, and the aerofoil will rise if the lift is sufficient to overcome the weight it carries.

Total lift depends on the type of aerofoil, the wing area, the speed of the plane and the density of the air. It is obviously difficult to get the plane off the ground and the engines must work at maximum force to create the necessary lift. Once the plane is airborne, however, less force is required to keep it in the air. To make the plane go higher, more speed is applied, because lift increases with velocity. It also increases according to the wing area. Because lift depends on pressures created by the air, it decreases at high altitudes when the air becomes thinner, or less dense.

INDEX

FORCES AND LAWS 1-4
FORCE, WORK AND WEIGHT 4-12
SIMPLE MACHINES 13-24
CONSERVATION OF ENERGY 25-29
MEASUREMENT 30-33
DYNAMICS AND MOVEMENT 33-40

Page numbers in italics refer to a diagram on that page.
Bold type refers to a heading or sub-heading.

A

Acceleration 6-8, *6*, 34, **37**, *37*
   gravity 6-7, *6*
Aerodynamics **39**, 40
Aerofoil 40, *40*
Aeroplane 1, *1*, 39-40
   lift *1*, 40, *40*
Air pressure 38
Area 30, **31**
Aristotle 2-3
Arm 14
Astronaut 5, *7*
Atoms 2
Axle 13, **15**, 16

B

Ball bearings 9
Bathysphere 39
Beam balance *5*
Bevel gears 22
Bicycle brakes 8
Block and tackle 16
Breadth 30

C

Calipers 30, *31*
Capstan 15, *16*, 17
Car jack 18, *19*
Centrifugal force **10**, 10-12, *10*, *11*
Centrifuge 11-12, *11*, *12*
Centripetal force **10**, 10-12, *11*
Chemical energy 25
Claw hammer *13*
Clay 38
Conservation of energy **29**
Copernicus 3
Crane 16, 17
Crow-bar *13*, 15
Cylinder (car) *9*

D

da Vinci, Leonardo 3
Deceleration 34, 37
Democritus 2
Density **31**, 32
Drag (aerodynamics) 39-40
Dynamics 33

E

Einstein, Albert 1, *2*
Electrical energy 28, 29, *29*

Electricity 29
Energy **25**, 25-29, *25*, *26*
   chemical 25
   conservation **29**
   heat 27, 29
   kinetic 27, 29
   potential 27
   electrical 28, 29

F

Falling bodies 3, *3*
Fishing rod 15
Flight **39**, 40
Force 1, **4**, *4*, 5
Friction **8**, 8-10, *9*, 14, 18, 27-28, *27*
Fulcrum 13, 14-15, *14*

G

Galileo 1, *1*, 3
Gases 38
Gearbox, 4-speed *23*
Gears **22**, 22-24, *22*, *23*
   compound train 24
   idler 22
   speed increaser 22
   speed reducer 22
Gravity 1, 5, *5*, **6**, 6-8, *6*, *7*, 8, 31-32, 37
   Earth 5-6, *5*, *7*
   moon 5-6, *5*

H

Heat energy 27
Helical gears 22
Hovercraft 9
Hydrofoil 9, *9*

I

Idler gear 22
Imperial measure 32
Inclined plane *6*, 13, **18**, *18*, *19*, 20
Inertia **10**

L

Leaning Tower of Pisa 3, *3*
Length 30
Lever 13, *13*, **14**, 14-15, *14*, 20

Lift (aerodynamics) *1*, 39-40, *40*
Lightning 29
Lubricants 8-9, *9*
Lubrication 8-9

M

Machines 1
   simple 13-25, *13*, *14*, *15*, *16*, *17*, *18*, *19*, *20*
Mass **5**, 5-6, **31**, 32
Matter 31
Measurement 30-33
   metric 32-33
Mechanical advantage **13**, 13-14, 16, 18
Mechanics 1
Metric system 32-33
Microscope 4
Motion 4
   first law *32*
   laws **33**, 34
   third law **36**, 37

N

Natural philosophy 2
Newton, Sir Isaac 1, *2*, 4
   laws of motion 6, *32*, 33-34, *36*, 37
Newton's Cradle *35*
Nutcracker *13*, 15

O

Oil 9, *9*

P

Parachute *8*
Pendulum *6*
Perpetual motion **27**, *27*, *28*, 29
Physics 2-4
   history **2**, 3
   modern **3**, 4
Pinion 22
Pitch (screw) 18
Pressure **37**, 38-40, *39*
   air 38
Pulley 13, *13*, **16**, 17, *17*
Pump-handle *14*
Pythagoras 2

R

Rack and pinion gears *22*, 24, *24*
Relativity, Theory of 1
Resistance *4*
Rest 4
Rotar *12*

S

Satellite *7, 8*
Scale 30
Screw 13, **18,** 18-20, *20, 21*
See-saw 15
Selector (gears) 24
Simple machines 13
    applications **20,** 20-22
Skateboarding *10,* 11
Sky-diver *8*
Space rocket *7, 36*
Spin dryer 12
Spiral staircase *21,* 22
Spring balance 6, 32
Spur gears *22*
Standard units **32**
Steps 22

T

Telescope *1,* 3, *4*
Thermodynamics, first law 28
Thickness **30**
Torque *15*

V

Velocity **34,** *34, 35,* 37
Velocity ratio 14
Vernier caliper 30
Volume *30,* **31**

W

Water 38
Water-wheel *15*
Wedge **20,** *20*
Weight **5,** 5-6, *5,* **31**
Wheel 13, **15,** 16
Wheelbarrow *14,* 15
Winch 17
Wind 9
Windlass *16,* 17
Wood 37-38
Work **4,** 13-14, **25,** 25-27